YOUR KNOWLEDGE HAS VALUE

- We will publish your bachelor's and master's thesis, essays and papers

- Your own eBook and book - sold worldwide in all relevant shops

- Earn money with each sale

Upload your text at www.GRIN.com and publish for free

Bibliographic information published by the German National Library:

The German National Library lists this publication in the National Bibliography; detailed bibliographic data are available on the Internet at http://dnb.dnb.de .

This book is copyright material and must not be copied, reproduced, transferred, distributed, leased, licensed or publicly performed or used in any way except as specifically permitted in writing by the publishers, as allowed under the terms and conditions under which it was purchased or as strictly permitted by applicable copyright law. Any unauthorized distribution or use of this text may be a direct infringement of the author s and publisher s rights and those responsible may be liable in law accordingly.

Imprint:

Copyright © 2016 GRIN Verlag, Open Publishing GmbH
Print and binding: Books on Demand GmbH, Norderstedt Germany
ISBN: 9783668371767

This book at GRIN:

http://www.grin.com/en/e-book/349889/occurrence-of-a-new-type-of-stingless-bee-in-kerala-evidence-from-morphometric

Prem Jose Vazhacharickal, Sajan Jose K.

Occurrence of a new type of stingless bee in Kerala. Evidence from morphometric analysis

GRIN Publishing

GRIN - Your knowledge has value

Since its foundation in 1998, GRIN has specialized in publishing academic texts by students, college teachers and other academics as e-book and printed book. The website www.grin.com is an ideal platform for presenting term papers, final papers, scientific essays, dissertations and specialist books.

Visit us on the internet:

http://www.grin.com/

http://www.facebook.com/grincom

http://www.twitter.com/grin_com

Occurrence of a new type of stingless bees in Kerala: Evidence from morphometric analysis

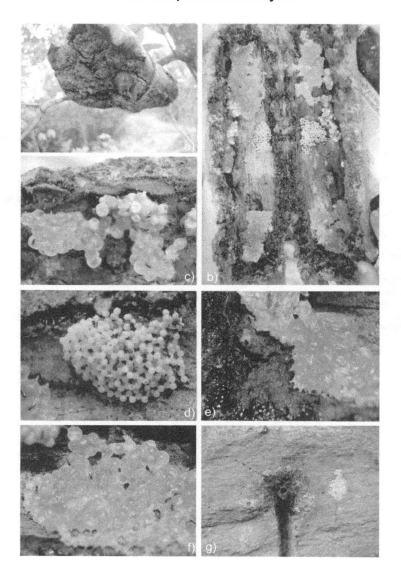

Prem Jose Vazhacharickal and Sajan Jose K

ACKNOWLEDGEMENT

Firstly we thank **God Almighty** whose blessing were always with us and helped us to complete this research work successfully.

The first author is extremely grateful to **Dr. Sajeshkumar N.K** (Head of the Department, Biotechnology) for the valuable suggestions, support and encouragements.

The first author would wish to thank beloved Manager **Rev. Fr. Dr. George Njarakunnel,** respected Principal **Dr. V.J. Joseph,** Bursar **Shaji Augustine,** Vice Principal **Fr. Joseph Allencheril**, and the Management for providing all the necessary facilities in carrying out the study.

We lovingly and gratefully indebted to our teachers, parents, siblings and friends who were there always for helping us in this project.

Prem Jose Vazhacharickal* and Sajan Jose K

*Address for correspondence
Assistant Professor
Department of Biotechnology
Mar Augusthinose College
Ramapuram-686576
Kerala, India
premjosev@gmail.com

Table of contents

Table of contents iii

Table of figures iv

Table of tables v

List of abbreviations vi

Occurrence of a new type of stingless bees in Kerala: Evidence from morphometric analysis 1

1. Introduction 2
2. Materials and Methods 3
   2.1 Study area 3
   2.2 Study design and data collection 3
   2.3 Morphometric measurements 4
   2.4 Statistical analysis 4
3. Results 5
   3.1 Descriptive analysis 5
   3.2 Correlation analysis 5
   3.3 Principal component analysis 5
4. Discussion 18
5. Conclusions 20
Acknowledgements 20
References 21

Table of figures

Figure 1. Morphometric characters considered for the analysis of worker and drone bees of stingless bees in Kerala: 1 = head length, 2 = head width, 3 = proboscis length, 4 = distance between two dorsal ocelli, 5 = dorsal ocello-ocular distance, 6 = antennal length, 7 = fermur length, 8 = tibia length, 9 = metatarsus length, 10 = metatarsus width, 11 = forewing length, 12 = forewing width, 13 = pterostigma length, 14 = pterostigma width, 15 = hind wing length, 16 = hind wing width, 17= number of hamuli, 18 = tergite length, 19 = sternite length, 20 = width of sternum, 21 = mandible length, 22 = mandible width. 6

Figure 2. Habitats of the newly reported stingless bees a) small entrance tube on laterite rock; b) breaking of laterite rock using iron wedges; c) cut opened laterite rock showing internal tunnel and brood; d) laterite rock part showing internal tunnel structure; e) honey pots; f) pollen pots; g) brood structure. 7

Figure 3. Habitats of the newly reported stingless bees in wooden log nest a) wooden log nest; b) nest opened showing internal structures; c) honey and pollen pots; d) bees with brood cells; e) and f) honey pots; g) entrance tube. 8

Figure 4. Scree plot for finding the various components to be extracted during Principal Component Analysis among stingless bee workers. 15

Figure 5. Component plot in rotated space during principal component analysis among stingless bee workers. 16

Figure 6. Distribution of factor scores among stingless bee workers across various locations in Kerala. 17

Table of tables

Table 1. Description of the morphometric variables used to classify *Trigona iridipennis* Smith in Kerala, India. 9

Table 2. Site description of the selected farms (D1-D9) across Kerala, India for collecting drone bees during 2011-2013. 10

Table 3. Description of *Trigona iridipennis* Smith worker and their morphometric characteristics. 10

Table 4. Description of some morphometric of *Trigona iridipennis* Smith worker and their characteristics. 10

Table 5. Pearson correlation coefficients between various morphometric characters of *Trigona iridipennis* Smith workers (n=30) in the various districts across Kerala, India during 2011-2013. 11

Table 6. Pearson correlation coefficients between various morphometric characters of *Trigona iridipennis* Smith workers (n=30) in the various districts across Kerala, India during 2011-2013. 12

Table 7. Eigenvalues and percentage of variance explained by the first four components in a PCA of *Trigona iridipennis* Smith worker in Kerala. 13

Table 8. Component matrix in a PCA of *Trigona iridipennis* Smith worker in Kerala. 13

Table 9. Rotated component matrix in a PCA of *Trigona iridipennis* Smith worker in Kerala. 14

Table 10. Component score coefficient matrix in a PCA of *Trigona iridipennis* Smith worker in Kerala. 14

List of abbreviations

AbSW	: Abdominal sternum width
AbTL	: Abdominal tergite length
AL	: Antennal length
ANOVA	: Analysis of variance
DBO	: Distance between two dorsal ocelli
DOOD	: Dorsal ocello-ocular distance
HAM	: Number of hamuli
HL	: Head length
HLW	: Head length width ratio
HW	: Head width
HWL	: Hind wing length
HWW	: Hind wing width
KS	: Kologovic-Smiroff
MdL	: Mandible length
MdW	: Mandible width
PCA	: Principal component analysis
PL	: Proboscis length
PtStW	: Pterostigma length
PtStW	: Pterostigma width
TFL	: Thorax femur length
TFWL	: Thorax forewing length
TFWLW	: Thorax forewing length width ratio
TFWW	: Thorax forewing width
TML	: Thorax metatarsus length
TMLW	: Thorax metatarsus length width ratio
TMW	: Thorax metatarsus width
TTL	: Thorax tibia length

Occurrence of a new type of stingless bees in Kerala: Evidence from morphometric analysis

Prem Jose Vazhacharickal[1]* and Sajan Jose K[2]

* premjosev@gmail.com

[1]Department of Biotechnology, Mar Augusthinose College, Ramapuram, Kerala, India-686576

[2]Department of Zoology, St. Joseph's College, Moolamattom, Kerala, India-685591

Abstract

Stingless bees are limited to tropics and subtropics with diversity in species and share morphological and behaviour patterns. Stingless bees are major pollinators of flowering plants in the tropics and improve crop productivity. *Trigona iridipennis* Smith are kept in India for centuries for the high medicinal value of honey as well as propolis and bee wax. A wide range of variations especially worker and drone body size and measurement were reported. Morphometric and geometric analysis provide a valuable tool for the discriminating variations among various honey bees and stingless bees. Based on these back ground, our objectives of this study were to 1) to characterize the morphometric aspects of workers in this reported new species of stingless bees 2) to identify the various similarities and differences existed among based on factor and principal component analysis with *Trigona iridipennis* Smith. A total of 30 samples of workers were collected and analyzed various morphometric characters including head length (HL), head width (HW), proboscis length (PL), distance between two dorsal ocelli/lower intercocular distance (DBO), dorsal ocello-ocular distance (DOOD), antennal length (AL), thorax femur length (TFL), thorax tibia length (TTL), thorax metatarsus length (TML), thorax metatarsus width (TMW), thorax forewing length (TFWL), thorax forewing width (TFWW), pterostigma width (PtStW), pterostigma length (PtStW), hind wing length (HWL), hind wing width (HWW), number of hamuli (HAM), abdominal tergite length (AbTL), abdominal sternum width (AbSW), mandible length (MdL) and mandible width (MdW), head length width ratio (HLW), thorax metatarsus length width ratio (TMLW) and thorax forewing length width ratio (TFWLW). One sample (SS1) from showed distinct differences in habitat preference (Laterite rock), appearance (size and colour), behaviour (passive) and nest architecture (narrow round entrance tube, snow white brood, cream white food pots). The present study based on the morphometry of stingless bees across Kerala shows that, in spite of the morphological and behavioural variations exhibited by the bees, members of all samples belong to *Trigona iridipennis*. But one sample (SS1) shows entirely different morphometric characteristics. It was found that this sample belongs to genus *Lisotrigona*

Keywords: Stingless bees; Meliponiculture; Cerumen; batumen; Brood.

1. Introduction

The decline in honey bee population happen due to destruction of natural habitat, intensive agriculture, diseases and climate change despite of role in pollination and ecosystem services (Hayes et al., 2008; Tomé et al., 2012; Neumann and Carreck, 2010; Cox-Foster et al., 2007; Ellis et al., 2010). Stingless bees are limited to tropics and subtropics with diversity in species and share morphological and behaviour patterns (Baumgartner and Roubik, 1989; Pignata et al., 1996). Stingless bees are major pollinators of flowering plants in the tropics and improve crop productivity (Ramanujam et al., 1993; Heard, 1999; Amano et al., 2000; Raju et al., 2009). The stingless bee found in Kerala is *Trigona iridipennis* Smith also called 'dammer bees' locally known as 'Cherutheneecha' in Malayalam (Singh, 2013).

Trigona iridipennis Smith are kept in India for centuries for the high medicinal value of honey as well as propolis and bee wax (Rasmussen, 2013; Cortopassi-Laurino et al., 2006; Virkar et al., 2014; Kumar et al., 2012; Andualem, 2013; Choudhari et al., 2013; Choudhari et al., 2012; Rasmussen, 2013) and serves a high demand from pharmaceutical sector (Kumar et al., 2012). Recent studies also showed the various nesting behaviour of *Trigona iridipennis* Smith in natural habitat as well as its adaptability various antroropoegnic habitats (Virkar et al., 2014; Singh, 2013; Nair and Nair, 2001; Kumar et al., 2012; Jose and Thomas, 2012; Jose and Thomas, 2013). A wide range of variations especially worker and drone body size and measurement were reported (Pignata et al., 1996; Vijayakumar and Jayraj, 2013; Danaraddi and Viraktamath, 2010; Danaraddi et al., 2012). Morphometric and geometric analysis provide a valuable tool for the discriminating variations among various honey bees and stingless bees (Vijayakumar and Jayraj, 2013; Wappler et al., 2012; Monteiro et al., 2002; Hernández et al., 2007; Francoy et al., 2011; Quezada-Euán et al., 2011; Hartfelder and Engels, 1992). These tools were sucessfuly utilized for the genetic lineage, geographic origin, heritability of shape, gender identification, differtiation and conservation (May-Itzá et al., 2012; May-Itzá et al., 2010; Quezada-Euán et al., 2011; Francoy et al., 2011; Francoy et al., 2009; Hernández et al., 2007; Monteiro et al., 2002; Rattanawannee et al., 2010; Combey et al., 2013). Further possibilities of phylogentic relationships were explored using advanced molecular biology tools especially DNA microsatellites, AFLP markers, mitochondrial DNA and nuclear DNA (Arias and Sheppard, 2005; Francisco et al.,

2008; Arias et al., 2006; Theeraapisakkun et al., 2010a; Theeraapisakkun et al., 2010b; Theeraapisakkun et al., 2011; May-Itzá et al., 2012). We report a new species of stingless bees in Kerala which are different from the morphometric studies of the *Trigona iridipennis* Smith. Based on these back ground, our objectives of this study were to 1) to characterize the morphometric aspects of workers in this reported new species of stingless bees 2) to identify the various similarities and differences existed among based on factor and principal component analysis with *Trigona iridipennis* Smith.

2. Materials and Methods

2.1 Study area

Kerala state covers an area of 38,863 km^2 with a population density of 859 per km^2 and spread across 14 districts. The climate is characterized by tropical wet and dry with average annual rainfall amounts to 2,817 ± 406 mm and mean annual temperature is 26.8°C (averages from 1871-2005; Krishnakumar et al., 2009). Maximum rainfall occurs from June to September mainly due to South West Monsoon and temperatures are highest in May and November.

2.2 Study design and data collection

Two locations (panchayaths) of these new stingless bees were located and collected based on farmer's information and initial baseline survey. A total of 10 worker samples were collected from these locations. The bees were caught using a self-designed net trap of size (200 μm) and the bees were immediately paralysed using chloroform solution (70%v/v) and later stored in eppendof vials with 70% alcohol. Non-destructive methods were followed using sample collection process. The samples were labelled according to the location and stored in airtight boxes until morphometric analysis were conducted. The positions of the locations were recorded with the help of a Trimble Geoexplorer II GPS (Trimble Navigation Ltd, Sunnyvale, CA, USA).

2.3 Morphometric measurements

The bees are dried at 60°C for 24 h in a hot air oven (Labtech 112, Iantech, Mumbai) to obtain their dry weight. After the weighing, the bees were rehydrated in glycerol solution (70%v/v) to soften their body. The head and thorax were dissected, and size of the following variables were recorded: head length (HL), head width (HW), proboscis length (PL), distance between two dorsal ocelli/lower inter-ocular distance (DBO), dorsal ocello-ocular distance (DOOD), antennal length (AL), thorax femur length (TFL), thorax tibia length (TTL), thorax metatarsus length (TML), thorax metatarsus width (TMW), thorax forewing length (TFWL), thorax forewing width (TFWW), pterostigma width (PtStW), pterostigma length (PtStW), hind wing length (HWL), hind wing width (HWW), number of hamuli (HAM), abdominal tergite length (AbTL), abdominal sternum width (AbSW), mandible length (MdL) and mandible width (MdW). Head length width ratio (HLW), thorax metatarsus length width ratio (TMLW) and thorax forewing length width ratio (TFWLW) were also determined. The morphometric measurements were made using steromicroscope (Leica E24D, Leica Microsystems, Switzerland) and analysed using LAS EZ version 1.4.0 (IBM Inc., CA, USA).

2.4 Statistical analysis

Descriptive statistics using SPSS 12.0 (SPSS Inc., Chicago, IL, USA) were conducted to summarize the data and graphs were generated using Sigma Plot 7 (Systat Software Inc., Chicago, IL, USA). The data were checked for normal distribution using Kologovic-Smiroff (KS) test. A comparison size metrics among the special samples collected from two different locations were compared with the usual samples collected across various districts and were analysed by means of Principal Component Analysis (PCA). Value scores for each worker and drone bees for each principal component were obtained from the first three components of PCA. Data were analysed by comparing the means at univariate level using ANOVA and multivariate approach (PCA). Colony scores from PCA were compared by means of ANOVA and plotted against pairs of factors.

A principal component analysis (PCA) was performed on the morphometric variables for the data reduction and obtain single measure of size (Wiley, 1981; Pignata and Diniz-Filho, 1996; Quezada-Euán et al., 2007). The PCA was carried out based on a correlation matrix (verified by means of Keiser-Meyer-Olkin measure of sampling adequacy and Bartlett's test of sphericity) and later analysed using VARIMAX axis rotation. The correlation matrix was selected as the derived principal component 1. A final measure of body size was given by the PC scores that were calculated for each individual as the product of the resulting coefficients in each PC (Quezada-Euán et al., 2007).

3. Results

3.1 Descriptive analysis

The cropping pattern of the collected sites vary with a wide variety plantation crops were natural rubber dominates all the other crops. The morphometric characters of workers varied widely across all the samples. Significant variations in the morphometric characters were observed for workers expect for PL; TML; MdW and AbTL. Highly significant difference ($P < 0.001$) were observed for HW; DBO; DOOD; AL; TFL; TTL; TMW; MdL; TFWL; TFWW; PtstW; PtstL; HWL and AbSW. The lowest HW among workers was 1.13 ± 0.04 (SS1) while height up to 1.60 ± 0.09 (S1). The lowest AL was seen in SS1 (1.36 ± 0.02) and highest in S1 (1.97 ± 0.04). The lowest MdL and MdW were 0.44 ± 0.01 and 0.16 ± 0.01 respectively. Even though numerical significance were observed in PL; TML; MdW and AbTL, but not able to prove their significance statistically.

3.2 Correlation analysis

Person correlation analysis showed considerable correlation among various morphometric characters in worker and drone bees across Kerala. Highly significant positive correlations were observed among HL and DL, HL and AL, HL and MdL, HL and TFL, HL and TTL, HW and DBO, HW and DOOB, HW and AL, HW and MdL, HW and MdW, HW and TFL, HW and TTL, DBO and AL, DBO and MdL, DBO and TFL, DBO and TTL, DOOB and AL, DOOB and TFL, DOOB and TTL, Al and MdW, Al and TFL, MdL and TFL, MdL and TTL, MdW TFL, MdW and TTL, TFL and TTL. Despite of strong positive correlations, strong negative correlation was observed among PL and DOOB, TMWL and PtStL, TML and PtStL, TWL and PtStL.

3.3 Principal component analysis

The validity PCA of 21 morphometric characters were justified as indicated by the Kaiser-Meyer-Olkin measure of sampling adequacy of 0.65. Bartlett's test of sphericity was significant (<0.001), indicating that the matrix was equally distributed and thus, suitable for PCA. The components were decided using scree plot and fixed to 4 components. The first three PCs explained 57.2, 12.3 and 10.1% of the total variation in the morphometric data respectively.

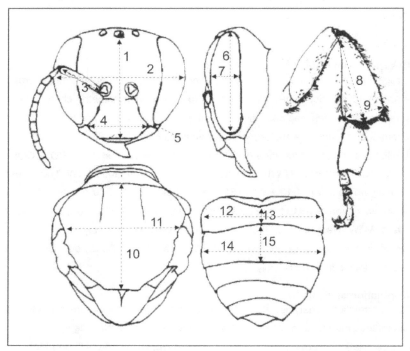

Figure 1. Morphometric characters considered for the analysis of worker and drone bees of stingless bees in Kerala: 1 = head length, 2 = head width, 3 = proboscis length, 4 = distance between two dorsal ocelli, 5 = dorsal ocello-ocular distance, 6 = antennal length, 7 = fermur length, 8 = tibia length, 9 = metatarsus length, 10 = metatarsus width, 11 = forewing length, 12 = forewing width, 13 = pterostigma length, 14 = pterostigma width, 15 = hind wing length, 16 = hind wing width, 17= number of hamuli, 18 = tergite length, 19 = sternite length, 20 = width of sternum, 21 = mandible length, 22 = mandible width.

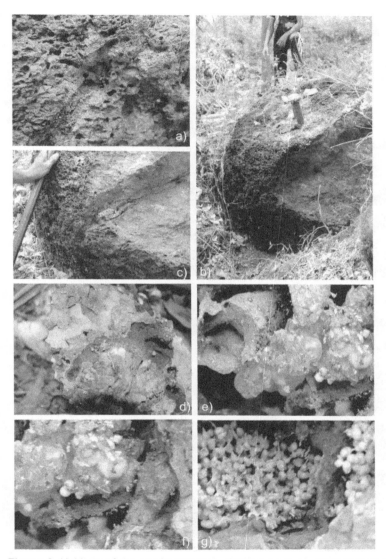

Figure 2. Habitats of the newly reported stingless bees a) small entrance tube on laterite rock; b) breaking of laterite rock using iron wedges; c) cut opened laterite rock showing internal tunnel and brood; d) laterite rock part showing internal tunnel structure; e) honey pots; f) pollen pots; g) brood structure.

Figure 3. Habitats of the newly reported stingless bees in wooden log nest a) wooden log nest; b) nest opened showing internal structures; c) honey and pollen pots; d) bees with brood cells; e) and f) honey pots; g) entrance tube.

Table 1. Description of the morphometric variables used to classify *Trigona iridipennis* Smith in Kerala, India.

Variable code	Description
HL	Length of head
HW	Width of head
HLW	Head length width ratio
PL	Proboscis length
DBO	Distance between two dorsal ocelli
DOOD	Dorsal ocello-ocular distance
AL	Antennal length
MdL	Mandible length
MdW	Mandible width
TFL	Thorax femur length
TTL	Thorax tibia length
TML	Thorax metatarsus length
TMW	Thorax metatarsus width
TMLW	Thorax metatarsus length width ratio
TFWL	Thorax forewing length
TFWW	Thorax forewing width
TFWLW	Thorax forewing length width ratio
PtStW	Pterostigma width
PtStL	Pterostigma length
HWL	Hind wing length
HWW	Hind wing width
HAM	Number of hamuli
AbTL	Abdominal tergite length
AbSL	Abdominal sternite length
AbSW	Abdominal sternite width

Table 2. Site description of the selected farms (S1, S2, SS1, S3) across Kerala, India for collecting drone bees during 2011-2013.

Site	Position	Type of hives	No. of samples	Crops
S1 (Sreekandapuram, Kannur)	12° 2' 21.58" N 75° 32' 42.50" E	Bamboo	10	Rubber, coconut
S2 (Kanjangad, Kasaragod)	12° 18' 47.98" N 75° 5' 45.24" E	Wooden box, areca nut	10	Rubber
SS1 (Sreekandapuram, Kannur)	12° 2' 21.58" N 75° 32' 42.50" E	Feral	10	Rubber, coconut
S3 (Kasaragod)	12° 26' 48.44" N 75° 17' 50.96" E	Feral	5	Rubber, areca nut

Table 3. Description of *Trigona iridipennis* Smith worker and their morphometric characteristics.

	HL	HW	PL	DBO	DOOD	AL	TFL	TTL	TML	TMW
Kannur (S1)	0.34 ± 0.03	1.60 ± 0.09	1.16 ± 0.09	0.38 ± 0.03	0.24 ± 0.03	1.97 ± 0.04	1.05 ± 0.05	1.48 ± 0.06	0.51 ± 0.05	0.28 ± 0.04
Kasaragod (S2)	0.33 ± 0.03	1.58 ± 0.12	1.19 ± 0.09	0.38 ± 0.01	0.20 ± 0.02	1.91 ± 0.05	1.04 ± 0.04	1.43 ± 0.05	0.54 ± 0.03	0.30 ± 0.03
Kannur (SS1)	0.29 ± 0.01	1.13 ± 0.04	1.13 ± 0.05	0.29 ± 0.01	0.17 ± 0.01	1.36 ± 0.02	0.65 ± 0.03	0.82 ± 0.01	0.38 ± 0.03	0.21 ± 0.01
Kasaragod (S3)	0.37 ± 0.01	1.56 ± 0.05	1.18 ± 0.03	0.37 ± 0.01	0.24 ± 0.01	1.96 ± 0.05	1.03 ± 0.02	1.41 ± 0.07	0.53 ± 0.03	0.28 ± 0.02
F values	5.631	30.528	0.622	22.307	8.938	207.791	114.050	160.041	15.647	8.848
P values	0.004	0.000	0.607	0.000	0.000	0.000	0.000	0.000	0.013	0.000

Numbers represent means ± one standard deviation (SD) of the mean.

Table 4. Description of some morphometric of *Trigona iridipennis* Smith worker and their characteristics.

	MdL	MdW	TFWL	TFWW	PtstW	PtstL	HWL	HWW	AbTL	AbSW
Kannur (S1)	0.54 ± 0.05	0.20 ± 0.01	3.65 ± 0.09	1.35 ± 0.05	0.55 ± 0.02	0.15 ± 0.02	2.58 ± 0.08	0.63 ± 0.03	0.49 ± 0.04	1.53 ± 0.09
Kasaragod (S2)	0.56 ± 0.04	0.19 ± 0.03	3.68 ± 0.07	1.37 ± 0.02	0.57 ± 0.02	0.16 ± 0.02	2.56 ± 0.08	0.65 ± 0.03	0.48 ± 0.03	1.45 ± 0.10
Kannur (SS1)	0.44 ± 0.01	0.16 ± 0.01	2.47 ± 0.05	0.96 ± 0.02	0.10 ± 0.01	0.37 ± 0.01	1.73 ± 0.03	0.53 ± 0.01	0.47 ± 0.01	1.28 ± 0.03
Kasaragod (S3)	0.51 ± 0.02	0.19 ± 0.02	3.57 ± 0.03	1.35 ± 0.03	0.13 ± 0.01	0.50 ± 0.02	2.47 ± 0.08	0.64 ± 0.01	0.48± 0.02	1.60 ± 0.06
F values	9.508	3.017	343.369	150.226	1056.492	392.767	169.435	24.348	0.401	13.187
P values	0.000	0.048	0.000	0.000	0.000	0.000	0.000	0.000	0.754	0.000

Numbers represent means ± one standard deviation (SD) of the mean.

Table 5. Pearson correlation coefficients between various morphometric characters of *Trigona iridipennis* Smith workers (n=35) in the various districts across Kerala, India during 2011-2013.

	HL	HW	PL	DBO	DOOB	AL	MdL	MdW	TFL	TTL
HL	1									
HW	0.361*	1								
PL	0.474**	0.163	1							
DBO	0.440*	0.742**	0.258	1						
DOOB	0.129	0.478**	-0.070	0.386*	1					
AL	0.544**	0.874**	0.249	0.811**	0.589**	1				
MdL	0.505**	0.537**	0.375*	0.745**	0.044	0.603**	1			
MdW	0.237	0.544**	0.184	0.459*	0.312	0.486**	0.242	1		
TFL	0.485**	0.841**	0.287	0.813**	0.508**	0.966*	0.605**	0.467**	1	
TTL	0.490**	0.885**	0.157	0.842**	0.535**	0.959*	0.657**	0.475**	0.953**	1

*Significant at the level of P< 0.05.

**Significant at the level of P< 0.01.

Table 6. Pearson correlation coefficients between various morphometric characters of *Trigona iridipennis* Smith workers (n=35) in the various districts across Kerala, India during 2011-2013.

	TML	TMW	TFWL	TFWW	PtStW	PtStL	HWL	HWW	AbTL	AbSL
TML	1									
TMW	0.762**	1								
TFWL	0.793**	0.730**	1							
TFWW	0.838**	0.742**	0.965**	1						
PtStW	0.512**	0.525**	0.713**	0.666**	1					
PtStL	-0.264	-0.360	-0.460**	-0.399*	0.934**	1				
HWL	0.790**	0.734**	0.976**	0.957**	0.738**	-0.495**	1			
HWW	0.821**	0.788**	0.845**	0.893**	0.577**	-0.334	0.866**	1		
AbTL	0.196	0.313	0.258	0.212	0.154	-0.148	0.293	0.297	1	
AbSL	0.351	0.342	0.438*	0.450*	-0.139	0.393*	0.446	0.524*	0.277	1

*Significant at the level of $P< 0.05$.

**Significant at the level of $P< 0.01$.

Table 7. Eigenvalues and percentage of variance explained by the first four components in a PCA of *Trigona iridipennis* Smith worker in Kerala.

Component	Eigenvalues	Percentage of variance	Cumulative percentage
1	9.157	57.234	57.234
2	1.975	12.345	69.579
3	1.630	10.188	79.768
4	0.843	5.266	85.034

Table 8. Component matrix in a PCA of *Trigona iridipennis* Smith worker in Kerala.

Character	Component			
	1	2	3	4
HL	0.502	0.642	0.337	0.281
HW	0.891	-0.256	0.145	-0.167
HLW	-0.517	0.716	0.164	0.350
PL	0.248	0.544	0.569	-0.217
DBO	0.906	0.119	-0.097	-0.113
DOOD	0.527	-0.456	0.150	0.526
AL	0.970	-0.019	0.068	0.109
MdL	0.696	0.463	-0.080	-0.270
MdW	0.509	-0.192	0.379	-0.374
TFL	0.955	0.007	0.039	0.031
TTL	0.968	-0.042	-0.004	0.034
TML	0.823	-0.108	-0.073	0.177
TMW	0.742	0.165	-0.578	0.077
TMLW	-0.068	-0.417	0.785	0.097
TFWL	0.983	0.016	-0.021	0.043
TFWW	0.976	0.024	-0.019	0.040

Table 9. Rotated component matrix in a PCA of *Trigona iridipennis* Smith worker in Kerala.

Character	Component			
	1	2	3	4
HL	0.462	0.761	-0.253	0.003
HW	0.786	0.074	0.527	-0.087
HLW	-0.454	0.515	-0.675	0.043
PL	0.045	0.829	0.184	-0.082
DBO	0.800	0.228	0.303	0.271
DOOD	0.704	-0.238	-0.107	-0.470
AL	0.935	0.198	0.212	0.002
MdL	0.513	0.488	0.259	0.459
MdW	0.332	0.209	0.618	-0.209
TFL	0.894	0.209	0.260	0.063
TTL	0.917	0.148	0.270	0.074
TML	0.843	0.011	0.113	0.030
TMW	0.751	-0.064	-0.023	0.591
TMLW	-0.058	0.092	0.202	-0.867
TFWL	0.928	0.186	0.242	0.114
TFWW	0.920	0.192	0.241	0.117

Table 10. Component score coefficient matrix in a PCA of *Trigona iridipennis* Smith worker in Kerala.

Character	Component			
	1	2	3	4
HL	0.110	0.344	-0.345	-0.115
HW	0.029	-0.008	0.264	-0.054
HLW	0.040	0.281	-0.483	-0.053
PL	-0.125	0.453	0.204	-0.049
DBO	0.045	0.049	0.111	0.134
DOOD	0.282	-0.196	-0.422	-0.398
AL	0.140	0.021	-0.069	-0.072
MdL	-0.059	0.211	0.207	0.274
MdW	-0.110	0.121	0.481	-0.070
TFL	0.107	0.032	0.002	-0.019
TTL	0.115	-0.003	0.003	-0.011
TML	0.165	-0.077	-0.145	-0.055
TMW	0.130	-0.131	-0.156	0.299
TMLW	0.011	0.092	0.070	-0.526
TFWL	0.118	0.014	-0.018	0.008
TFWW	0.115	0.018	-0.016	0.010

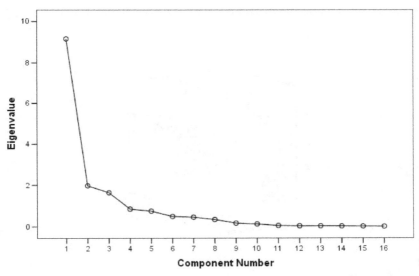

Figure 4. Scree plot for finding the various components to be extracted during Principal Component Analysis among stingless bee workers.

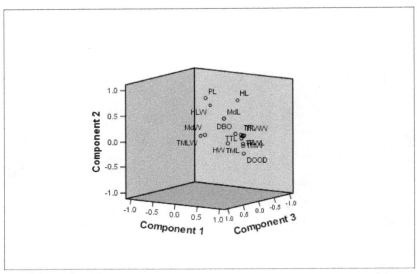

Figure 5. Component plot in rotated space during principal component analysis among stingless bee workers.

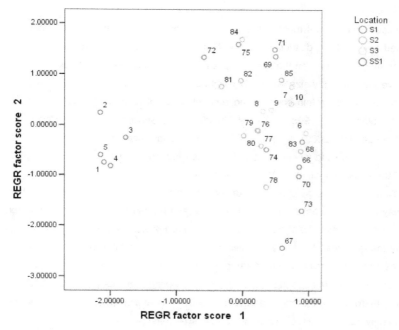

Figure 6. Distribution of factor scores among stingless bee workers across various locations in Kerala.

4. Discussion

Our studies revealed that most of the worker morphometric characters were positively correlated across all the districts in Kerala. The morphometric characters investigated in our study mainly depend on the body on body size as well as weight. Previous studies also prove that the body size and body weight of stingless bees were highly correlated and depend on the protein content and amount of larval food (Quezada-Euán et al., 2011; Ramalho et al., 1998; Hilário et al., 2003; Del Sarto et al., 2005; Hilário ans Imperatriz-Fonseca, 2009). In addition the worker body size vary depending on the season of brood raising where 10 fold differences in worker sizes observed in bumble bees (**Quezada-Euán et al., 2011**; Couvillon and Dornhaus, 2010). Considering the typical climatic conditions, availability of foraging plants and nature of domestication, there may exist significant changes in body size and related characteristics. However, there is not much significant differences among drones across various districts. This may due the facts that the drone samples were not fully representative of all districts of Kerala as well the special nature of sample collection emphasising on non-destructive method of collection. In adopting non-destructive sampling methods especially for drone collection, there exit some technical issues especially its availability. The drones will be available only during aggressive swarming as well as mating with gyne.

Positive correlation for majority of morphometric characters among worker bees also support the linkage of body size which may relate to the feed quality, availability and season. This can be much cleared understood in the contest of geographic positions were some of the sample collection areas were on sea side but others on plain as well as high altitude with cold climate.

Protein content variation of larval food were already reported across various seasons among stingless bees as well as in other honey bee species (Quezada-Euán et al., 2011; Brodschneider and Crailsheim, 2010; Haydak, 1970; Roulston and Cane, 2000; Hartfelder and Engels, 1989; Roulston et al., 2000). The usage more sugar in larval food in absence of pollen source can also relate to the size difference, where larvae fed with more sugar become much smaller. This phenomenal is present in bumble bees which make the colony much competent in resources usage efficiency as well as foraging efficiency (Quezada-Euán et al., 2011; Goulson et al., 2002; Goulson and Sparrow 2009; Ings et al., 2005; Couvillon and Dornhaus, 2010).

The production of the brood decreases during rainy seasons and dearth although the colones have strong food reserve. These also suggest the brood cells are provisioned based on the incoming pollen as well as its availability (Quezada-Euán et al., 2011; Roubik, 1982; Moo-Valle et al., 2001; Hilário and Imperatriz-Fonseca, 2009; Peat et al., 2005; Pech-May et al., 2012). The availability of pollen differ according to seasons as well as area and its quality depend on the protein content and digestibility (Quezada-Euán et al., 2011; Keller et al., 2005; Robertson et al., 1999; Hoover et al., 2006; Roulston and Cane, 2000; Standifer et al., 1980).

Previous studies on morphometry of stingless bees reveal some differences among populations (Danaraddi and Viraktamath, 2010; Danaraddi et al., 2010; Rassmussen, 2013). The morphometric studies and the comparison of the samples from all districts in Kerala with the previous work show that irrespective of the variations members of all the populations can be treated as *Trigona iridipennis*. The present study agrees with Devanesan et al. (2003), reported the stingless bees in Kerala as *Trigona iridipennis*. One sample (SS1) showed distinct differences in habitat preference (Laterite rock), appearance (size and colour), behaviour (passive) and nest architecture (narrow round entrance tube, snow white brood, cream white food pots). It was found that SS1 shows significant difference in all the characters compared to other samples, but shows similarity with the one reported by Viraktmath et al. (2013). Morphometric comparison of this sample with other workers (Engel, 2000; Jobiraj and Narendran, 2004; Chinh et al., 2005; Rasmussen, 2013) and with Viraktamath et al., 2013 clearly points out that, this sample belong to the genus *Lisotrigona*. But it is yet to be named, as it demand further verification. These tiny stingless bees have also been reported from Sri Lanka, India, Vietnam, Sumatra and China (Rasmussen, 2008). May be recent molecular biology tools especially with nuclear and mitochondrial DNA may further benefited to differentiate the various species of stingless bees in Kerala.

5. Conclusions

The present study based on the morphometry of stingless bees across Kerala shows that, in spite of the morphological and behavioural variations exhibited by the bees, members of all samples belong to *Trigona iridipennis*. But one sample (SS1) shows entirely different morphometric characteristics. It was found that this sample belongs to genus *Lisotrigona*. It demands further investigations to establish its species level identity. Thus, this study resulted in reporting the occurrence of two genera of stingless bees in Kerala. The most abundant and widely distributed one is *Trigona iridipennis*, and other is an unnamed one in the genus *Lisotrigona*. However advanced molecular biology studies also merit further research to support our research.

Acknowledgements

The funding from UGC and KSCSTE is highly appreciated. The analytical support form University of Agricultural Sciences, Dharwad is acknowledged. We also thank all the farmers who cooperated with us during the study.

References

Amano, K., Nemoto, T., & Heard, T. A. (2000). What are stingless bees, and why and how to use them as crop pollinators?-a review. *Japan Agricultural Research Quarterly*, *34*(3), 183-190.

Andualem, B. (2013). Synergistic Antimicrobial effect of Tenegn honey (*Trigona iridipennis*) and garlic against standard and clinical pathogenic Bacterial isolates. *International Journal of Microbiological Research*, *4*(1), 16-22.

Arias, M. C., & Sheppard, W. S. (2005). Phylogenetic relationships of honey bees (Hymenoptera: Apinae: Apini) inferred from nuclear and mitochondrial DNA sequence data. *Molecular Phylogenetics and Evolution*, *37*(1), 25-35.

Arias, M. C., Magalhães, R., & Flávio de Oliveira, F. (2006). Molecular markers as a tool for population and evolutionary studies of stingless bees. *Apidologie*, *37*(1), 259-274.

Bänziger, H., Pumikong, S., & Srimuang, K. O. (2011). The remarkable nest entrance of tear drinking *Pariotrigona klossi* and other stingless bees nesting in limestone cavities (Hymenoptera: Apidae). *Journal of the Kansas Entomological Society*, *84*(1), 22-35.

Basavarajappa, S. (2010). Studies on the impact of anthropogenic interference on wild honeybees in Mysore District, Karnataka, India. *African Journal of Agricultural Research*, *5*(4), 298-305.

Baumgartner, D. L., & Roubik, D. W. (1989). Ecology of necrophilous and filth-gathering stingless bees (Apidae: Meliponinae) of Peru. *Journal of the Kansas Entomological Society*, *62*(1), 11-22.

Brodschneider, R., & Crailsheim, K. (2010). Nutrition and health in honey bees. *Apidologie*, *41*(3), 278-294.

Cameron, E. C., Franck, P., & Oldroyd, B. P. (2004). Genetic structure of nest aggregations and drone congregations of the southeast Asian stingless bee Trigona collina. *Molecular Ecology*, *13*(8), 2357-2364.

Chinh, T. X., Sommeijer, M. J., Boot, W. J., & Michener, C. D. (2005). Nest and colony characteristics of three stingless bee species in Vietnam with the first description of the nest of *Lisotrigona carpenteri* (Hymenoptera: Apidae: Meliponini). *Journal of the Kansas Entomological Society, 78*(4), 363-372.

Choudhari, M. K., Haghniaz, R., Rajwade, J. M., & Paknikar, K. M. (2013). Anticancer activity of Indian stingless bee propolis: an in vitro study. *Evidence-Based Complementary and Alternative Medicine, 2013*(1), 1-10.

Choudhari, M. K., Punekar, S. A., Ranade, R. V., & Paknikar, K. M. (2012). Antimicrobial activity of stingless bee (*Trigona sp.*) propolis used in the folk medicine of Western Maharashtra, India. *Journal of Ethnopharmacology, 141*(1), 363-367.

Combey, R., Teixeira, J. S. G., Bonatti, V., Kwapong, P., & Francoy, T. M. (2013). Geometric morphometrics reveals morphological differentiation within four African stingless bee species. *Annals of Biological Research, 4*(11), 93-103.

Cortopassi-Laurino, M., Imperatriz-Fonseca, V. L., Roubik, D. W., Dollin, A., Heard, T., Aguilar, I., & Nogueira-Neto, P. (2006). Global meliponiculture: challenges and opportunities. *Apidologie, 37*(2), 275-292.

Couvillon, M. J., & Dornhaus, A. (2010). Small worker bumble bees (Bombus impatiens) are hardier against starvation than their larger sisters. *Insectes Sociaux, 57*(2), 193-197.

Couvillon, M. J., Jandt, J. M., Duong, N. H. I., & Dornhaus, A. (2010). Ontogeny of worker body size distribution in bumble bee (Bombus impatiens) colonies. *Ecological Entomology, 35*(4), 424-435.

Cox-Foster, D. L., Conlan, S., Holmes, E. C., Palacios, G., Evans, J. D., Moran, N. A., & Lipkin, W. I. (2007). A metagenomic survey of microbes in honey bee colony collapse disorder. *Science, 318*(5848), 283-287.

Danaraddi, C. S., & Viraktamath, S. (2007). Studies on stingless bee, *Trigona iridipennis* Smith with special reference to foraging behaviour and melissopalynology at Dharwad, Karnataka. *Master of Science Thesis. College of Agricultural Science. Dharwad.*

Danaraddi, C. S., & Viraktamath, S. (2010). Morphometrical studies on the stingless bee, *Trigona iridipennis* Smith. *Karnataka Journal of Agricultural Sciences, 22*(4), 796-797.

Danaraddi, C. S., Hakkalappanavar, S., Biradar, S. B., Tattimani, M., & Dandagi, M. R. (2012). Morphometrical studies on the stingless bee, *Trigona iridipennis* Smith. *Asian Journal of Biological Sciences and Technology*, 7(1), 49-51.

Danaraddi, C. S., Viraktamath, S., Basavanagoud, K., & Bhat, A. R. S. (2010). Nesting habits and nest structure of stingless bee, *Trigona iridipennis* Smith at Dharwad, Karnataka. *Karnataka Journal of Agricultural Sciences*, 22(2), 310-313.

David, W. R. (2006). Stingless bee nesting biology. Apidologie, 37, 124-143.

Del Sarto, M. C. L., Peruquetti, R. C., & Campos, L. A. O. (2005). Evaluation of the neotropical stingless bee *Melipona quadrifasciata* (Hymenoptera: Apidae) as pollinator of greenhouse tomatoes. *Journal of Economic Entomology*, 98(2), 260-266.

Devanesan, S., Shailaja, K. K., & Pramila, K. S. (2009). Status paper on Stingless bee *Trigona iridipennis* Smith. All India Co-ordinated Research Project on Honey bees and pollinators, Vellayani Centre,Thiruvananthapuram. pp79.

Ellis, J. D., Evans, J. D., & Pettis, J. (2010). Colony losses, managed colony population decline, and Colony Collapse Disorder in the United States. *Journal of Apicultural Research*, 49(1), 134-136.

Engel, M. S. (2000). Classification of the bee tribe Augochlorini (Hymenoptera: Halictidae). *Bulletin of the American Museum of Natural History*, 250(1), 1-89.

Francisco, F. O., Nunes-Silva, P., Francoy, T. M., Wittmann, D., Imperatriz-Fonseca, V. L., Arias, M. C., & Morgan, E. D. (2008). Morphometrical, biochemical and molecular tools for assessing biodiversity. An example in *Plebeia remota* (Holmberg, 1903)(Apidae, Meliponini). *Insectes Sociaux*, 55(3), 231-237.

Franck, P., Cameron, E., Good, G., Rasplus, J. Y., & Oldroyd, B. P. (2004). Nest architecture and genetic differentiation in a species complex of Australian stingless bees. *Molecular Ecology*, 13(8), 2317-2331.

Francoy, T. M., Grassi, M. L., Imperatriz-Fonseca, V. L., de Jesús May-Itzá, W., & Quezada-Euán, J. J. G. (2011). Geometric morphometrics of the wing as a tool for assigning genetic lineages and geographic origin to *Melipona beecheii* (Hymenoptera: Meliponini). *Apidologie*, 42(4), 499-507.

Francoy, T. M., Silva, R. A. O., Nunes-Silva, P., Menezes, C., & Imperatriz-Fonseca, V. L. (2009). Gender identification of five genera of stingless bees (Apidae, Meliponini) based on wing morphology. *Genetics and Molecular Research, 8*(1), 207-214.

Goulson, D., & Sparrow, K. R. (2009). Evidence for competition between honeybees and bumblebees; effects on bumblebee worker size. *Journal of Insect Conservation, 13*(2), 177-181.

Goulson, D., Peat, J., Stout, J. C., Tucker, J., Darvill, B., Derwent, L. C., & Hughes, W. O. (2002). Can alloethism in workers of the bumblebee, *Bombus terrestris*, be explained in terms of foraging efficiency?. *Animal Behaviour, 64*(1), 123-130.

Hartfelder, K., & Engels, W. (1989). The composition of larval food in stingless bees: evaluating nutritional balance by chemosystematic methods. *Insectes Sociaux, 36*(1), 1-14.

Hartfelder, K., & Engels, W. (1989). The composition of larval food in stingless bees: evaluating nutritional balance by chemosystematic methods. *Insectes Sociaux, 36*(1), 1-14.

Hartfelder, K., & Engels, W. (1992). Allometric and multivariate analysis of sex and caste polymorphism in the neotropical stingless bee, *Scaptotrigona postica*. *Insectes Sociaux, 39*(3), 251-266.

Hartfelder, K., & Engels, W. (1992). Allometric and multivariate analysis of sex and caste polymorphism in the neotropical stingless bee, *Scaptotrigona postica*. *Insectes Sociaux, 39*(3), 251-266.

Haydak, M. H. (1970). Honey bee nutrition. *Annual review of entomology, 15*(1), 143-156.

Hayes Jr, J., Underwood, R. M., & Pettis, J. (2008). A survey of honey bee colony losses in the US, fall 2007 to spring 2008. *PloS One, 3*(12), e4071.

Heard, T. A. (1999). The role of stingless bees in crop pollination. *Annual Review of Entomology, 44*(1), 183-206.

Hernández, E. J., Roubik, D. W., & Nates-Parra, G. (2007). Morphometric analysis of bees in the *Trigona fulviventris* group (Hymenoptera: Apidae). *Journal of the Kansas Entomological Society, 80*(3), 205-212.

Hilário, S. D., & Imperatriz-Fonseca, V. L. (2009). Pollen foraging in colonies of *Melipona bicolor* (Apidae, Meliponini): effects of season, colony size and queen number. *Genetics and Molecular Research*, *8*(2), 664-671.

Hilário, S. D., Gimenes, M., & Imperatriz-Fonseca, V. L. (2003). The influence of colony size in diel rhythms of flight activity of *Melipona bicolor* Lepeletier (Hymenoptera, Apidae, Meliponini). *Apoidea neotropica: Homenagem aos*, *90*, 191-197.

Hoover, S. E., Higo, H. A., & Winston, M. L. (2006). Worker honey bee ovary development: seasonal variation and the influence of larval and adult nutrition. *Journal of Comparative Physiology B*, *176*(1), 55-63.

Ings, T. C., Schikora, J., & Chittka, L. (2005). Bumblebees, humble pollinators or assiduous invaders? A population comparison of foraging performance in *Bombus terrestris*. *Oecologia*, *144*(3), 508-516.

Jaenike, J., & Holt, R. D. (1991). Genetic variation for habitat preference: evidence and explanations. *American Naturalist*, *137*(1), S67-S90.

Jobiraj, T., & Narendran, T. C. (2004). A revised key to the world species of *Lisotrigona moure* (Hymenoptera: Apoidea: Apidae) with description of a new species from India. *Entomon*, *29*, 39-44.

Jose, S. K & Thomas, S. (2012). Stingless beekeeping (Meliponiculture) in Kerala. In: Selected beneficial and harmful insects of Indian subcontinent, (Thomas, K.S (20102) ed. LAP LAMBERT Academic Publishing GmBH & Co. KG, Saarbruken, Germany.

Jose, S. K., & Thomas, S. (2013). Nest architecture of *Trigona iridipennis*. Proceedings of the National Seminar on Invertebrate Taxonomy. Nirmala Academic and Research Publications, Kerala, India.

Keller, I., Fluri, P., & Imdorf, A. (2005). Pollen nutrition and colony development in honey bees: part I. *Bee World*, *86*(1), 3-10.

Krishnakumar, K. N., Prasada Rao, G. S. L. H. V., & Gopakumar, C. S. (2009). Rainfall trends in twentieth century over Kerala, India. *Atmospheric Environment*, *43*(11), 1940-1944.

Kumar, M. S., Singh, A. J. A. R., & Alagumuthu, G. (2012). Traditional beekeeping of stingless bee (Trigona sp) by Kani tribes of Western Ghats, Tamil Nadu, India. *Indian Journal of Traditional Knowledge, 11*(2), 342-345.

Lima, S. L., & Dill, L. M. (1990). Behavioral decisions made under the risk of predation: a review and prospectus. *Canadian Journal of Zoology, 68*(4), 619-640.

Marisa, H., & Salni, S. (2012). Red wood (*Pterocarpus indicus* wild) and bread fruit (*Artocarpus communis*) bark sap as attractant of stingless bee (*Trigona spp*). *Malaysian Journal of Fundamental and Applied Sciences, 8*(2), 107-110.

Martin, T. E. (2001). Abiotic vs. biotic influences on habitat selection of coexisting species: climate change impacts?. *Ecology, 82*(1), 175-188.

May-Itzá, W. D. J., Quezada-Euán, J. J. G., Ayala, R., & De La Rúa, P. (2012). Morphometric and genetic analyses differentiate Mesoamerican populations of the endangered stingless bee Melipona beecheii (Hymenoptera: Meliponidae) and support their conservation as two separate units. *Journal of Insect Conservation, 16*(5), 723-731.

May-Itzá, W. D. J., Quezada-Euán, J. J. G., Medina, L. A. M., Enríquez, E., & De la Rúa, P. (2010). Morphometric and genetic differentiation in isolated populations of the endangered Mesoamerican stingless bee *Melipona yucatanica* (Hymenoptera: Apoidea) suggest the existence of a two species complex. *Conservation Genetics, 11*(5), 2079-2084.

Michener, C. D. (1974). *The social behaviour of the bees*. A comparative study. Belknap Press, Cambridge, UK.

Mohan, R., & Devanesan, S. (1999). Dammer bees, *Trigona iridipennis* Smith.*Apidae: Meliponinae) in Kerala. Insect Environment, 5*(2), 79.

Monteiro, L. R., Diniz-Filho, J. A. F., Reis, S. F., & Araújo, E. D. (2002). Geometric estimates of heritability in biological shape. *Evolution, 56*(3), 563-572.

Moo-Valle, H., Quezada-Euán, J. J., & Wenseleers, T. (2001). The effect of food reserves on the production of sexual offspring in the stingless bee *Melipona beecheii* (Apidae, Meliponini). *Insectes Sociaux, 48*(4), 398-403.

Nair, M. C. & Nair, P. K. K. (2001) Beekeeping by Kanikkars in southern Western Ghats of Kerala. *Indian Bee Journal, 63*(1), 11-16.

Nair, M. C. (2003). Apiculture resource biodiversity and management in Southern Kerala. PhD. Thesis. Mahatma Ghandhi University, Kottayam, 277.

Neumann, P., & Carreck, N. L. (2010). Honey bee colony losses. *Journal of Apicultural Research, 49*(1), 1-6.

Pavithra, N., Shankar, R., Jayaprakash. (2012). Nesting pattern preferences of stingless bee, *Trigona iridipennis* Smith (Hymenoptera: Apidae) in Jnanabharathi campus, Karnataka, India. *International Research Journal of Biological Sciences, 2*(2), 44-50.

Peat, J., Darvill, B., Ellis, J., & Goulson, D. (2005). Effects of climate on intra-and interspecific size variation in bumble-bees. *Functional Ecology, 19*(1), 145-151.

Pech-May, F. G., Medina-Medina, L., May-Itzá, W. D. J., Paxton, R. J., & Quezada-Euán, J. J. G. (2012). Colony pollen reserves affect body size, sperm production and sexual development in males of the stingless bee *Melipona beecheii. Insectes Sociaux, 59*(3), 417-424.

Pignata, M. I. B., & Diniz-Filho, J. A. F. (1996). Phylogenetic autocorrelation and evolutionary constraints in worker body size of some neotropical stingless bees (Hymenoptera: Apidae). *Heredity, 76*(3), 222-228.

Quezada-Euán, J. J. G., López-Velasco, A., Pérez-Balam, J., Moo-Valle, H., Velazquez-Madrazo, A., & Paxton, R. J. (2011). Body size differs in workers produced across time and is associated with variation in the quantity and composition of larval food in Nannotrigona perilampoides (Hymenoptera, Meliponini). *Insectes sociaux, 58*(1), 31-38.

Quezada-Euán, J. J. G., Paxton, R. J., Palmer, K. A., Itzá, W. D. J. M., Tay, W. T., & Oldroyd, B. P. (2007). Morphological and molecular characters reveal differentiation in a Neotropical social bee, Melipona beecheii (Apidae: Meliponini). *Apidologie, 38*(3), 247-258.

Raju, A. J. S., Rao, K. S., & Rao, N. G. (2009). Association of Indian Stingless Bee, *Trigona iridipennis* Smith (Apidae: Meliponinae) with Red-listed *Cycas sphaerica* Roxb.(Cycadaceae). *Current Science, 96*(11), 1435-1436.

Ramalho, M., Imperatriz-Fonseca, V. L., & Giannini, T. C. (1998). Within-colony size variation of foragers and pollen load capacity in the stingless bee Melipona quadrifasciata anthidioides Lepeletier (Apidae, Hymenoptera). *Apidologie, 29*(3), 221-228.

Ramanujam, C. G. K., Fatima, K., & Kalpana, T. P. (1993). Nectar and pollen sources for dammer bee (*Trigona iridipennis* Smith) in Hyderabad (India). *Indian Bee Journal, 55*(1/2), 25-28.

Rasmussen, C. (2013). Stingless bees (Hymenoptera: Apidae: Meliponini) of the Indian subcontinent: Diversity, taxonomy and current status of knowledge. *Zootaxa, 3647*(3), 401-428.

Rattanawannee, A., Chanchao, C., & Wongsiri, S. (2010). Gender and species identification of four native honey bees (Apidae: Apis) in Thailand based on wing morphometic analysis. *Annals of the Entomological Society of America, 103*(6), 965-970.

Robertson, A. W., Mountjoy, C., Faulkner, B. E., Roberts, M. V., & Macnair, M. R. (1999). Bumble bee selection of *Mimulus guttatus* flowers: the effects of pollen quality and reward depletion. *Ecology, 80*(8), 2594-2606.

Roubik, D. W. (1982). Seasonality in colony food storage, brood production and adult survivorship: studies of Melipona in tropical forest (Hymenoptera: Apidae).*Journal of the Kansas Entomological Society, 56*(1), 789-800.

Roulston, T. A. H., Cane, J. H., & Buchmann, S. L. (2000). What governs protein content of pollen: pollinator preferences, pollen-pistil interactions, or phylogeny?. *Ecological Monographs, 70*(4), 617-643.

Roulston, T. H., & Cane, J. H. (2000). Pollen nutritional content and digestibility for animals. *Plant Systematics and Evolution, 222*(1-4), 187-209.

Singh, R. P. (2013). Domestication of *Trigona iridipennis* Smith in a newly designed hive. *National Academy Science Letters, 36*(4), 367-371.

Slaa, E. J., Chaves, L. A. S., Malagodi-Braga, K. S., & Hofstede, F. E. (2006). Stingless bees in applied pollination: practice and perspectives. *Apidologie, 37*(2), 293-315.

Standifer, L. N., McCaughey, W. F., Dixon, S. E., GILLIAM, M., & Loper, G. M. (1980). Biochemistry and microbiology of pollen collected by honey bees (*Apis mellifera* l.) from almond, *Prunus dulcis*. ii. protein, amino acids and enzymes (1). *Apidologie*, *11*(2), 163-171.

Theeraapisakkun, M., Klinbunga, S., & Sittipraneed, S. (2010a). Development of a species-diagnostic marker and its application for population genetics studies of the stingless bee *Trigona collina* in Thailand. *Genetics and Molecular Research*, *9*(2), 919-930.

Thummajitsakul, S., Klinbunga, S., & Sittipraneed, S. (2010b). Development of a species-diagnostic marker for identification of the stingless bee Trigona pagdeni in Thailand. *Biochemical Genetics*, *48*(3-4), 181-192.

Thummajitsakul, S., Klinbunga, S., & Sittipraneed, S. (2011). Genetic differentiation of the stingless bee *Tetragonula pagdeni* in Thailand using SSCP analysis of a large subunit of mitochondrial ribosomal DNA. *Biochemical Genetics*, *49*(7-8), 499-510.

Tomé, H. V. V., Martins, G. F., Lima, M. A. P., Campos, L. A. O., & Guedes, R. N. C. (2012). Imidacloprid-induced impairment of mushroom bodies and behavior of the native stingless bee Melipona quadrifasciata anthidioides. *PloS One*, *7*(6), e38406.

Vijayakumar, K., & Jayaraj, R. Geometric morphometry analysis of three species of stingless bees in India. *International Journal for Life Science and Educational Research*, *1*(2), 91-95.

Viraktamath, S., Rajankar, B., Jose, S. K., & Thomas, S. (2013). *Lisotrigona* species of stingless bees. In: Morphometry and phylogeography of honey bees and stingless bees in India. 218-225.

Virkar, P. S., Shrotriya, S., & Uniyal, V. P. (2014). Building Walkways: Observation on Nest Duplication of Stingless Bee *Trigona iridipennins* Smith (1854). *Ambient Science*, *1*(1), 38-40.

Wappler, T., De Meulemeester, T., Murat Aytekin, A., Michez, D., & Engel, M. S. (2012). Geometric morphometric analysis of a new Miocene bumble bee from the Randeck Maar of southwestern Germany (Hymenoptera: Apidae). *Systematic Entomology*, *37*(4), 784-792.

Wiley, E. O., & Lieberman, B. S. (2011). *Phylogenetics: theory and practice of phylogenetic systematics.* John Wiley & Sons, New York, USA.

YOUR KNOWLEDGE HAS VALUE

- We will publish your bachelor's and master's thesis, essays and papers

- Your own eBook and book - sold worldwide in all relevant shops

- Earn money with each sale

Upload your text at www.GRIN.com
and publish for free

Printed in the USA
CPSIA information can be obtained
at www.ICGtesting.com
LVHW051928170924
791294LV00003B/600